THIS BOOK BELONGS TO

Name :

....

....

....

....

Humble Math

Table of Contents

Sections : Day :

Adding Digits 0-5 1-7

Adding Digits 0-7 8-14

Adding Digits 0-10 15-35

Subtraction Digits 0-10 15-42

Subtraction Digits 10-20 43-56

Subtraction Digits 0-20 57-70

(Answer Key in Back)

+ −

Day 1
Adding digits 0-5

Time : ____ : ____ minutes **Score :** ________ out of 36

1. 5 + 5 =	2. 0 + 0 =	3. 4 + 5 =	4. 0 + 2 =	5. 5 + 1 =	6. 4 + 5 =
7. 5 + 1 =	8. 4 + 5 =	9. 0 + 4 =	10. 3 + 0 =	11. 1 + 1 =	12. 1 + 4 =
13. 2 + 3 =	14. 2 + 3 =	15. 2 + 3 =	16. 4 + 0 =	17. 5 + 0 =	18. 5 + 2 =
19. 0 + 5 =	20. 4 + 1 =	21. 0 + 1 =	22. 0 + 4 =	23. 0 + 0 =	24. 5 + 1 =
25. 0 + 2 =	26. 1 + 5 =	27. 2 + 2 =	28. 3 + 2 =	29. 2 + 2 =	30. 3 + 1 =
31. 5 + 0 =	32. 2 + 5 =	33. 2 + 5 =	34. 1 + 2 =	35. 4 + 5 =	36. 3 + 1 =

Day 2
Adding digits 0-5

Time : ____ : ____ minutes **Score :** ______ out of 36

1	2	3	4	5	6
3 + 1 =	0 + 1 =	1 + 0 =	4 + 0 =	4 + 0 =	5 + 1 =
7	8	9	10	11	12
3 + 3 =	3 + 2 =	0 + 3 =	1 + 1 =	4 + 0 =	0 + 2 =
13	14	15	16	17	18
2 + 4 =	5 + 5 =	1 + 4 =	1 + 4 =	0 + 1 =	3 + 3 =
19	20	21	22	23	24
2 + 0 =	4 + 1 =	0 + 5 =	2 + 1 =	1 + 3 =	0 + 0 =
25	26	27	28	29	30
4 + 1 =	2 + 5 =	5 + 3 =	4 + 4 =	2 + 1 =	0 + 3 =
31	32	33	34	35	36
0 + 0 =	4 + 0 =	0 + 4 =	4 + 1 =	0 + 4 =	4 + 3 =

Day 3
Adding digits 0-5

Time : ____:____ minutes **Score :** ________ out of 36

1	2	3	4	5	6
1 + 5	2 + 3	1 + 1	0 + 0	1 + 0	1 + 1

7	8	9	10	11	12
0 + 3	0 + 4	2 + 3	2 + 0	1 + 0	2 + 5

13	14	15	16	17	18
4 + 1	3 + 0	2 + 1	4 + 4	3 + 3	3 + 1

19	20	21	22	23	24
1 + 4	1 + 4	1 + 0	4 + 2	5 + 5	0 + 0

25	26	27	28	29	30
3 + 4	0 + 2	2 + 1	3 + 5	5 + 0	4 + 2

31	32	33	34	35	36
3 + 2	2 + 3	0 + 2	0 + 3	5 + 3	0 + 4

Day 4
Adding digits 0-5

Time : ____ : ____ minutes **Score :** ______ out of 36

1	2	3	4	5	6
3 + 2 =	1 + 2 =	4 + 2 =	4 + 1 =	0 + 2 =	3 + 4 =

7	8	9	10	11	12
2 + 5 =	4 + 5 =	2 + 0 =	5 + 3 =	2 + 3 =	2 + 1 =

13	14	15	16	17	18
0 + 3 =	2 + 0 =	5 + 5 =	5 + 4 =	1 + 3 =	1 + 0 =

19	20	21	22	23	24
0 + 5 =	4 + 4 =	5 + 1 =	0 + 3 =	4 + 2 =	1 + 1 =

25	26	27	28	29	30
2 + 0 =	3 + 4 =	3 + 5 =	3 + 1 =	0 + 2 =	0 + 3 =

31	32	33	34	35	36
2 + 3 =	5 + 4 =	3 + 4 =	5 + 5 =	3 + 1 =	2 + 1 =

Day 5
Adding digits 0-5

Time : ______ : ______ minutes **Score :** ________ out of 36

1	2	3	4	5	6
1 + 4	0 + 1	3 + 3	1 + 4	1 + 1	1 + 1
7	8	9	10	11	12
2 + 1	0 + 4	3 + 1	5 + 0	2 + 1	5 + 3
13	14	15	16	17	18
0 + 1	5 + 3	4 + 0	3 + 2	5 + 0	3 + 3
19	20	21	22	23	24
3 + 0	0 + 1	5 + 0	2 + 0	5 + 4	1 + 1
25	26	27	28	29	30
3 + 5	0 + 3	0 + 5	0 + 0	1 + 5	3 + 2
31	32	33	34	35	36
2 + 4	2 + 3	2 + 0	5 + 4	3 + 5	4 + 4

Day 6
Adding digits 0-5

Time : ____ : ____ minutes **Score :** ________ out of 36

1	2	3	4	5	6
2 + 3	2 + 1	0 + 0	0 + 0	4 + 4	1 + 1

7	8	9	10	11	12
4 + 1	2 + 5	4 + 4	3 + 0	5 + 4	3 + 3

13	14	15	16	17	18
1 + 1	2 + 3	5 + 2	2 + 1	3 + 1	1 + 5

19	20	21	22	23	24
0 + 0	1 + 5	5 + 0	1 + 0	2 + 5	0 + 1

25	26	27	28	29	30
2 + 4	2 + 5	2 + 2	2 + 1	4 + 2	0 + 0

31	32	33	34	35	36
0 + 3	5 + 4	3 + 0	2 + 3	5 + 0	0 + 3

Day 7
Adding digits 0-5

Time : ____ : ____ minutes **Score :** ________ out of 36

1	2	3	4	5	6
$5 + 4 =$	$2 + 3 =$	$0 + 2 =$	$3 + 4 =$	$1 + 4 =$	$3 + 4 =$
7	8	9	10	11	12
$1 + 1 =$	$2 + 1 =$	$2 + 2 =$	$5 + 0 =$	$1 + 5 =$	$3 + 5 =$
13	14	15	16	17	18
$5 + 2 =$	$5 + 2 =$	$3 + 5 =$	$4 + 4 =$	$3 + 0 =$	$5 + 4 =$
19	20	21	22	23	24
$5 + 4 =$	$1 + 4 =$	$0 + 4 =$	$1 + 1 =$	$3 + 2 =$	$2 + 5 =$
25	26	27	28	29	30
$1 + 3 =$	$3 + 1 =$	$1 + 5 =$	$5 + 5 =$	$2 + 1 =$	$1 + 4 =$
31	32	33	34	35	36
$2 + 1 =$	$5 + 4 =$	$3 + 2 =$	$5 + 0 =$	$2 + 4 =$	$0 + 4 =$

Day 8
Adding digits 0-7

Time : ______ : ______ minutes **Score :** ________ out of 36

1. 5 + 0 =	2. 4 + 7 =	3. 1 + 1 =	4. 6 + 1 =	5. 7 + 0 =	6. 1 + 3 =
7. 5 + 6 =	8. 5 + 6 =	9. 4 + 5 =	10. 3 + 6 =	11. 0 + 3 =	12. 6 + 4 =
13. 6 + 7 =	14. 3 + 3 =	15. 3 + 4 =	16. 3 + 1 =	17. 6 + 1 =	18. 4 + 5 =
19. 3 + 5 =	20. 3 + 6 =	21. 0 + 5 =	22. 3 + 2 =	23. 5 + 7 =	24. 2 + 7 =
25. 3 + 2 =	26. 3 + 0 =	27. 6 + 3 =	28. 5 + 2 =	29. 4 + 0 =	30. 4 + 4 =
31. 3 + 7 =	32. 0 + 6 =	33. 1 + 6 =	34. 1 + 3 =	35. 0 + 2 =	36. 6 + 3 =

Day 9
Adding digits 0-7

Time : ____:____ minutes **Score :** ________ out of 36

1	2	3	4	5	6
0 + 5	1 + 2	5 + 6	4 + 6	2 + 3	1 + 4

7	8	9	10	11	12
6 + 3	2 + 6	6 + 5	0 + 2	5 + 3	1 + 6

13	14	15	16	17	18
0 + 5	3 + 0	7 + 3	1 + 5	2 + 4	3 + 4

19	20	21	22	23	24
6 + 6	6 + 1	6 + 2	0 + 2	3 + 3	7 + 3

25	26	27	28	29	30
5 + 0	4 + 6	1 + 6	7 + 7	6 + 5	0 + 2

31	32	33	34	35	36
3 + 0	1 + 4	5 + 4	6 + 6	5 + 6	6 + 0

Day 10
Adding digits 0-7

Time : ______ : ______ minutes **Score :** ________ out of 36

1	2	3	4	5	6
2 + 2 =	6 + 7 =	2 + 6 =	6 + 5 =	0 + 6 =	4 + 7 =
7	8	9	10	11	12
3 + 5 =	2 + 4 =	6 + 3 =	7 + 4 =	0 + 7 =	1 + 3 =
13	14	15	16	17	18
6 + 3 =	0 + 4 =	5 + 4 =	1 + 0 =	4 + 6 =	3 + 6 =
19	20	21	22	23	24
0 + 3 =	7 + 5 =	5 + 4 =	3 + 3 =	0 + 3 =	1 + 1 =
25	26	27	28	29	30
7 + 3 =	4 + 2 =	5 + 6 =	0 + 7 =	5 + 0 =	5 + 7 =
31	32	33	34	35	36
5 + 2 =	3 + 0 =	0 + 3 =	7 + 4 =	0 + 1 =	4 + 1 =

Day 11
Adding digits 0-7

Time : ____ : ____ minutes **Score :** ________ out of 36

1 7 + 7	2 4 + 0	3 3 + 4	4 6 + 1	5 2 + 5	6 3 + 0
7 2 + 4	8 7 + 4	9 7 + 0	10 2 + 4	11 6 + 3	12 6 + 2
13 5 + 1	14 7 + 3	15 0 + 1	16 1 + 1	17 5 + 4	18 5 + 2
19 0 + 5	20 1 + 4	21 0 + 6	22 4 + 7	23 2 + 2	24 5 + 3
25 0 + 7	26 2 + 2	27 0 + 2	28 2 + 4	29 7 + 1	30 6 + 7
31 1 + 3	32 6 + 7	33 1 + 6	34 7 + 5	35 2 + 6	36 5 + 2

Day 12
Adding digits 0-7

Time : ____:____ minutes **Score :** ________ out of 36

1	2	3	4	5	6
4 + 1 =	0 + 3 =	4 + 4 =	1 + 1 =	4 + 1 =	3 + 7 =

7	8	9	10	11	12
1 + 2 =	2 + 3 =	5 + 6 =	2 + 0 =	6 + 1 =	4 + 4 =

13	14	15	16	17	18
4 + 5 =	7 + 1 =	1 + 4 =	3 + 5 =	5 + 7 =	7 + 3 =

19	20	21	22	23	24
0 + 4 =	7 + 0 =	3 + 0 =	7 + 6 =	0 + 3 =	7 + 3 =

25	26	27	28	29	30
4 + 1 =	0 + 0 =	0 + 5 =	5 + 7 =	0 + 0 =	0 + 7 =

31	32	33	34	35	36
0 + 3 =	6 + 0 =	3 + 2 =	3 + 5 =	7 + 4 =	6 + 3 =

Day 13
Adding digits 0-7

Time : ______ : ______ minutes **Score :** ________ out of 36

1	2	3	4	5	6
0 + 6	0 + 2	4 + 1	1 + 5	0 + 0	5 + 1
7	8	9	10	11	12
5 + 7	3 + 3	4 + 7	1 + 1	3 + 2	1 + 7
13	14	15	16	17	18
6 + 6	1 + 2	4 + 7	5 + 0	7 + 5	0 + 3
19	20	21	22	23	24
0 + 0	0 + 0	3 + 4	4 + 0	2 + 4	6 + 2
25	26	27	28	29	30
3 + 4	1 + 7	6 + 5	3 + 0	6 + 3	4 + 4
31	32	33	34	35	36
3 + 3	0 + 5	1 + 4	5 + 7	5 + 2	2 + 6

Day 14

Adding digits 0-7

Time : ______ : ______ minutes **Score :** ________ out of 36

1 3 + 7	2 6 + 4	3 2 + 3	4 2 + 5	5 7 + 6	6 2 + 1
7 6 + 2	8 6 + 1	9 6 + 2	10 0 + 0	11 1 + 6	12 2 + 6
13 4 + 0	14 5 + 0	15 6 + 1	16 1 + 3	17 5 + 6	18 3 + 7
19 3 + 1	20 0 + 0	21 4 + 0	22 1 + 5	23 4 + 6	24 2 + 2
25 4 + 4	26 1 + 2	27 4 + 2	28 6 + 6	29 1 + 4	30 5 + 4
31 4 + 2	32 6 + 7	33 1 + 5	34 2 + 5	35 1 + 1	36 0 + 7

Day 15
Adding digits 0-10

Time : ____:____ minutes **Score :** ________ out of 36

1. 9 + 1 =	2. 2 + 7 =	3. 2 + 8 =	4. 6 + 8 =	5. 5 + 1 =	6. 2 + 8 =
7. 1 + 3 =	8. 9 + 2 =	9. 4 + 2 =	10. 10 + 5 =	11. 5 + 10 =	12. 0 + 9 =
13. 0 + 4 =	14. 10 + 8 =	15. 10 + 3 =	16. 5 + 5 =	17. 0 + 1 =	18. 4 + 6 =
19. 10 + 9 =	20. 1 + 1 =	21. 9 + 3 =	22. 5 + 10 =	23. 3 + 2 =	24. 6 + 1 =
25. 6 + 6 =	26. 8 + 9 =	27. 7 + 4 =	28. 10 + 5 =	29. 0 + 10 =	30. 10 + 7 =
31. 0 + 1 =	32. 3 + 1 =	33. 3 + 6 =	34. 1 + 9 =	35. 8 + 7 =	36. 5 + 3 =

Day 16
Adding digits 0-10

Time : ______ : ______ minutes **Score :** ________ out of 36

1	2	3	4	5	6
5 + 2	3 + 3	1 + 2	0 + 5	3 + 6	8 + 10
7	8	9	10	11	12
7 + 3	5 + 0	7 + 0	6 + 6	7 + 2	3 + 10
13	14	15	16	17	18
9 + 7	5 + 2	0 + 2	7 + 1	2 + 2	3 + 4
19	20	21	22	23	24
9 + 7	3 + 5	3 + 5	0 + 7	8 + 2	5 + 2
25	26	27	28	29	30
1 + 0	0 + 3	3 + 5	6 + 9	10 + 1	9 + 2
31	32	33	34	35	36
0 + 7	8 + 9	3 + 7	6 + 4	6 + 8	5 + 2

Day 17
Adding digits 0-10

Time : ____:____ minutes **Score :** ________ out of 36

1	2	3	4	5	6
0 + 10 =	7 + 8 =	6 + 0 =	9 + 5 =	1 + 7 =	4 + 0 =

7	8	9	10	11	12
8 + 4 =	5 + 2 =	8 + 6 =	2 + 2 =	10 + 0 =	4 + 4 =

13	14	15	16	17	18
10 + 4 =	1 + 6 =	7 + 1 =	0 + 0 =	1 + 9 =	4 + 10 =

19	20	21	22	23	24
4 + 7 =	9 + 0 =	3 + 6 =	8 + 8 =	5 + 7 =	2 + 3 =

25	26	27	28	29	30
2 + 8 =	2 + 1 =	2 + 2 =	1 + 5 =	8 + 5 =	1 + 10 =

31	32	33	34	35	36
5 + 10 =	7 + 7 =	10 + 0 =	2 + 7 =	3 + 5 =	3 + 0 =

Day 18
Adding digits 0-10

Time : ____ : ____ minutes **Score :** ______ out of 36

1	2	3	4	5	6
4 + 1	6 + 2	9 + 8	4 + 0	9 + 4	6 + 10
7	8	9	10	11	12
7 + 9	9 + 8	2 + 5	5 + 4	5 + 10	0 + 2
13	14	15	16	17	18
7 + 10	5 + 6	9 + 3	5 + 4	2 + 6	1 + 7
19	20	21	22	23	24
2 + 0	5 + 7	8 + 8	1 + 7	10 + 0	2 + 3
25	26	27	28	29	30
4 + 3	1 + 2	0 + 4	8 + 0	7 + 1	1 + 3
31	32	33	34	35	36
2 + 6	10 + 10	6 + 1	7 + 6	5 + 0	8 + 10

Day 19
Adding digits 0-10

Time : ____:____ minutes **Score :** ______ out of 36

1. 8 + 10	2. 3 + 4	3. 1 + 5	4. 0 + 2	5. 0 + 2	6. 4 + 2
7. 8 + 5	8. 2 + 5	9. 3 + 10	10. 3 + 7	11. 1 + 8	12. 10 + 4
13. 4 + 8	14. 0 + 3	15. 7 + 0	16. 5 + 1	17. 9 + 6	18. 2 + 8
19. 7 + 4	20. 6 + 2	21. 8 + 8	22. 9 + 9	23. 2 + 1	24. 1 + 1
25. 10 + 9	26. 9 + 3	27. 8 + 2	28. 0 + 9	29. 1 + 8	30. 9 + 1
31. 0 + 1	32. 8 + 2	33. 10 + 10	34. 8 + 0	35. 4 + 9	36. 8 + 6

Day 20
Adding digits 0-10

Time : ____ : ____ minutes **Score :** ______ out of 36

1. 7 + 3	2. 6 + 9	3. 1 + 3	4. 7 + 0	5. 7 + 7	6. 6 + 0
7. 10 + 3	8. 4 + 2	9. 10 + 3	10. 8 + 4	11. 5 + 5	12. 3 + 8
13. 10 + 9	14. 2 + 0	15. 4 + 2	16. 5 + 10	17. 1 + 0	18. 7 + 8
19. 6 + 0	20. 2 + 8	21. 1 + 0	22. 2 + 6	23. 10 + 6	24. 4 + 9
25. 7 + 7	26. 4 + 9	27. 4 + 3	28. 9 + 6	29. 6 + 10	30. 2 + 3
31. 4 + 9	32. 10 + 4	33. 8 + 1	34. 5 + 7	35. 2 + 1	36. 10 + 0

Day 21
Adding digits 0-10

Time : ______ : ______ minutes **Score :** ________ out of 36

1 8 + 5	2 4 + 4	3 1 + 8	4 9 + 1	5 2 + 0	6 4 + 7
7 3 + 8	8 9 + 0	9 6 + 3	10 9 + 1	11 5 + 1	12 8 + 6
13 2 + 9	14 8 + 8	15 3 + 8	16 0 + 7	17 8 + 0	18 0 + 8
19 3 + 1	20 8 + 8	21 7 + 2	22 2 + 8	23 4 + 8	24 8 + 6
25 4 + 3	26 8 + 6	27 6 + 5	28 9 + 4	29 6 + 3	30 8 + 0
31 7 + 9	32 10 + 1	33 5 + 1	34 0 + 4	35 5 + 2	36 8 + 3

Day 22
Adding digits 0-10

Time : ______ : ______ minutes **Score :** ________ out of 36

1. 9 + 4 =	2. 0 + 3 =	3. 5 + 8 =	4. 1 + 7 =	5. 8 + 0 =	6. 7 + 4 =
7. 3 + 8 =	8. 7 + 9 =	9. 5 + 8 =	10. 5 + 5 =	11. 7 + 9 =	12. 7 + 6 =
13. 2 + 4 =	14. 1 + 8 =	15. 10 + 5 =	16. 5 + 7 =	17. 5 + 3 =	18. 2 + 8 =
19. 6 + 1 =	20. 2 + 2 =	21. 2 + 8 =	22. 4 + 1 =	23. 1 + 8 =	24. 0 + 2 =
25. 2 + 7 =	26. 6 + 6 =	27. 4 + 1 =	28. 6 + 6 =	29. 7 + 7 =	30. 4 + 8 =
31. 3 + 6 =	32. 1 + 4 =	33. 3 + 2 =	34. 9 + 5 =	35. 8 + 10 =	36. 1 + 6 =

Day 23
Adding digits 0-10

Time : ____ : ____ minutes **Score :** ________ out of 36

1	2	3	4	5	6
6 + 6	10 + 0	0 + 6	6 + 6	3 + 5	10 + 6

7	8	9	10	11	12
7 + 7	1 + 3	3 + 3	1 + 7	5 + 8	6 + 10

13	14	15	16	17	18
4 + 9	8 + 2	2 + 3	6 + 6	3 + 8	4 + 7

19	20	21	22	23	24
1 + 10	5 + 8	3 + 10	4 + 5	1 + 3	1 + 1

25	26	27	28	29	30
8 + 6	1 + 2	7 + 3	10 + 5	10 + 7	5 + 4

31	32	33	34	35	36
7 + 0	2 + 3	5 + 10	5 + 2	0 + 9	7 + 0

Day 24
Adding digits 0-10

Time : ______ : ______ minutes **Score :** ________ out of 36

1. 0 + 1 =	2. 8 + 6 =	3. 6 + 2 =	4. 9 + 0 =	5. 8 + 3 =	6. 1 + 0 =
7. 6 + 6 =	8. 2 + 0 =	9. 10 + 7 =	10. 7 + 10 =	11. 9 + 3 =	12. 10 + 5 =
13. 6 + 1 =	14. 8 + 2 =	15. 8 + 10 =	16. 0 + 2 =	17. 5 + 9 =	18. 2 + 9 =
19. 5 + 3 =	20. 2 + 10 =	21. 1 + 2 =	22. 0 + 2 =	23. 5 + 4 =	24. 1 + 4 =
25. 10 + 6 =	26. 9 + 3 =	27. 7 + 10 =	28. 3 + 10 =	29. 9 + 6 =	30. 0 + 3 =
31. 5 + 3 =	32. 0 + 4 =	33. 9 + 5 =	34. 5 + 5 =	35. 3 + 5 =	36. 2 + 8 =

Day 25
Adding digits 0-10

Time : ____ : ____ minutes **Score :** ________ out of 36

1	2	3	4	5	6
8 + 7	3 + 0	7 + 0	5 + 3	1 + 4	3 + 9

7	8	9	10	11	12
6 + 10	1 + 8	2 + 5	8 + 9	10 + 3	3 + 5

13	14	15	16	17	18
5 + 10	9 + 5	7 + 4	5 + 9	0 + 7	7 + 9

19	20	21	22	23	24
10 + 7	9 + 5	10 + 1	0 + 2	3 + 1	1 + 4

25	26	27	28	29	30
1 + 4	3 + 3	0 + 2	5 + 6	2 + 6	8 + 0

31	32	33	34	35	36
0 + 5	0 + 6	7 + 8	9 + 8	10 + 6	8 + 10

Day 26
Adding digits 0-10

Time : ____ : ____ minutes **Score :** ______ out of 36

1	2	3	4	5	6
8 + 9 =	2 + 10 =	9 + 0 =	3 + 2 =	9 + 4 =	5 + 2 =

7	8	9	10	11	12
3 + 8 =	3 + 9 =	3 + 4 =	4 + 8 =	8 + 3 =	8 + 3 =

13	14	15	16	17	18
9 + 8 =	3 + 4 =	1 + 5 =	1 + 8 =	5 + 7 =	3 + 0 =

19	20	21	22	23	24
5 + 10 =	8 + 1 =	9 + 8 =	10 + 2 =	1 + 4 =	2 + 5 =

25	26	27	28	29	30
10 + 0 =	0 + 0 =	0 + 5 =	8 + 4 =	3 + 10 =	2 + 7 =

31	32	33	34	35	36
1 + 2 =	2 + 5 =	7 + 1 =	10 + 7 =	4 + 4 =	1 + 6 =

Day 27
Adding digits 0-10

Time : ______ : ______ minutes **Score :** ________ out of 36

1	2	3	4	5	6
8 + 7 =	8 + 0 =	8 + 0 =	9 + 6 =	6 + 4 =	10 + 3 =

7	8	9	10	11	12
4 + 8 =	9 + 9 =	10 + 8 =	1 + 8 =	0 + 0 =	6 + 3 =

13	14	15	16	17	18
7 + 4 =	2 + 8 =	9 + 7 =	6 + 3 =	2 + 10 =	8 + 6 =

19	20	21	22	23	24
9 + 2 =	2 + 9 =	6 + 4 =	9 + 8 =	0 + 4 =	6 + 7 =

25	26	27	28	29	30
3 + 6 =	8 + 10 =	9 + 7 =	8 + 8 =	3 + 7 =	7 + 4 =

31	32	33	34	35	36
3 + 3 =	6 + 7 =	3 + 1 =	6 + 5 =	4 + 3 =	7 + 5 =

Day 28
Adding digits 0-10

Time : ____ : ____ minutes **Score :** ________ out of 36

1. 6 + 5 =	2. 9 + 9 =	3. 4 + 4 =	4. 5 + 8 =	5. 3 + 2 =	6. 4 + 3 =
7. 10 + 2 =	8. 10 + 2 =	9. 0 + 1 =	10. 0 + 3 =	11. 1 + 6 =	12. 5 + 0 =
13. 9 + 10 =	14. 4 + 4 =	15. 4 + 7 =	16. 5 + 2 =	17. 2 + 7 =	18. 8 + 7 =
19. 1 + 6 =	20. 6 + 1 =	21. 1 + 7 =	22. 3 + 6 =	23. 2 + 1 =	24. 3 + 2 =
25. 4 + 8 =	26. 8 + 2 =	27. 1 + 4 =	28. 6 + 3 =	29. 1 + 1 =	30. 4 + 8 =
31. 4 + 4 =	32. 2 + 2 =	33. 4 + 2 =	34. 3 + 10 =	35. 0 + 4 =	36. 2 + 6 =

Day 29
Adding digits 0-10

Time : ____:____ minutes **Score :** ________ out of 36

1 0 + 1	2 4 + 6	3 4 + 5	4 4 + 8	5 1 + 10	6 3 + 3
7 10 + 6	8 5 + 5	9 5 + 9	10 9 + 1	11 1 + 2	12 5 + 4
13 6 + 1	14 4 + 5	15 9 + 2	16 7 + 2	17 3 + 0	18 1 + 6
19 4 + 6	20 4 + 8	21 4 + 1	22 9 + 7	23 10 + 10	24 9 + 7
25 8 + 8	26 0 + 7	27 3 + 2	28 2 + 1	29 10 + 5	30 5 + 7
31 2 + 1	32 4 + 9	33 7 + 4	34 5 + 6	35 7 + 6	36 10 + 5

Day 30
Adding digits 0-10

Time : ____ : ____ minutes **Score :** ________ out of 36

1	2	3	4	5	6
2 + 6 =	8 + 0 =	6 + 2 =	3 + 4 =	6 + 4 =	7 + 3 =

7	8	9	10	11	12
6 + 3 =	0 + 5 =	7 + 1 =	5 + 6 =	8 + 6 =	7 + 5 =

13	14	15	16	17	18
2 + 3 =	3 + 2 =	10 + 3 =	10 + 1 =	6 + 2 =	2 + 4 =

19	20	21	22	23	24
4 + 10 =	5 + 9 =	8 + 7 =	7 + 6 =	4 + 6 =	4 + 0 =

25	26	27	28	29	30
0 + 10 =	1 + 4 =	1 + 0 =	2 + 5 =	8 + 6 =	3 + 3 =

31	32	33	34	35	36
6 + 0 =	9 + 7 =	7 + 9 =	1 + 1 =	2 + 0 =	7 + 6 =

Day 31
Adding digits 0-10

Time : ____:____ minutes **Score :** ________ out of 36

1 6 + 0	2 2 + 4	3 10 + 0	4 6 + 4	5 9 + 5	6 0 + 6
7 7 + 5	8 4 + 7	9 10 + 6	10 8 + 9	11 10 + 7	12 3 + 3
13 7 + 2	14 4 + 6	15 6 + 2	16 2 + 8	17 5 + 6	18 4 + 6
19 0 + 6	20 3 + 8	21 5 + 0	22 1 + 5	23 7 + 3	24 1 + 0
25 9 + 10	26 10 + 0	27 5 + 8	28 1 + 1	29 8 + 10	30 9 + 5
31 2 + 10	32 9 + 1	33 1 + 5	34 10 + 5	35 1 + 2	36 5 + 1

Day 32
Adding digits 0-10

Time : ____:____ minutes **Score :** ________ out of 36

1	2	3	4	5	6
5 + 10	2 + 5	8 + 7	9 + 10	7 + 7	4 + 7
7	**8**	**9**	**10**	**11**	**12**
9 + 7	4 + 8	1 + 0	1 + 6	3 + 10	2 + 3
13	**14**	**15**	**16**	**17**	**18**
2 + 10	10 + 0	1 + 6	3 + 6	7 + 4	7 + 3
19	**20**	**21**	**22**	**23**	**24**
4 + 9	0 + 9	9 + 0	0 + 9	6 + 2	0 + 6
25	**26**	**27**	**28**	**29**	**30**
2 + 8	9 + 5	2 + 4	8 + 5	9 + 1	8 + 0
31	**32**	**33**	**34**	**35**	**36**
1 + 7	5 + 2	6 + 2	10 + 1	0 + 5	6 + 6

Day 33
Adding digits 0-10

Time : ____ : ____ minutes **Score :** ________ out of 36

1. 10 + 3 =	2. 4 + 8 =	3. 4 + 0 =	4. 1 + 5 =	5. 10 + 0 =	6. 4 + 5 =
7. 7 + 0 =	8. 2 + 8 =	9. 8 + 7 =	10. 0 + 9 =	11. 8 + 8 =	12. 6 + 0 =
13. 3 + 0 =	14. 8 + 8 =	15. 8 + 2 =	16. 1 + 6 =	17. 2 + 9 =	18. 1 + 9 =
19. 3 + 1 =	20. 9 + 10 =	21. 1 + 9 =	22. 2 + 10 =	23. 6 + 3 =	24. 10 + 0 =
25. 1 + 1 =	26. 1 + 5 =	27. 6 + 8 =	28. 4 + 5 =	29. 5 + 7 =	30. 8 + 8 =
31. 9 + 3 =	32. 9 + 6 =	33. 1 + 10 =	34. 6 + 2 =	35. 4 + 4 =	36. 5 + 8 =

Day 34
Adding digits 0-10

Time : ____ : ____ minutes **Score :** ______ out of 36

1	2	3	4	5	6
9 + 5	0 + 6	2 + 5	7 + 2	1 + 5	7 + 3

7	8	9	10	11	12
7 + 4	0 + 8	10 + 5	2 + 1	6 + 7	5 + 1

13	14	15	16	17	18
7 + 5	7 + 8	9 + 8	4 + 7	7 + 2	6 + 2

19	20	21	22	23	24
3 + 7	10 + 8	10 + 10	9 + 2	2 + 7	5 + 2

25	26	27	28	29	30
9 + 7	7 + 10	5 + 4	9 + 1	4 + 8	4 + 9

31	32	33	34	35	36
1 + 3	9 + 2	8 + 10	2 + 2	5 + 7	8 + 0

Day 35
Adding digits 0-10

Time : ____ : ____ minutes **Score :** ________ out of 36

1	2	3	4	5	6
5 + 0 =	6 + 8 =	0 + 6 =	5 + 6 =	1 + 7 =	6 + 1 =

7	8	9	10	11	12
5 + 1 =	4 + 0 =	5 + 1 =	3 + 5 =	4 + 8 =	5 + 8 =

13	14	15	16	17	18
4 + 9 =	1 + 4 =	8 + 0 =	6 + 3 =	8 + 8 =	10 + 1 =

19	20	21	22	23	24
0 + 7 =	8 + 1 =	6 + 7 =	2 + 4 =	2 + 7 =	7 + 3 =

25	26	27	28	29	30
2 + 0 =	5 + 2 =	2 + 10 =	4 + 2 =	8 + 3 =	5 + 2 =

31	32	33	34	35	36
8 + 6 =	9 + 2 =	3 + 2 =	7 + 4 =	10 + 3 =	2 + 7 =

Day 36

Subtraction digits 0-10

Time : ______ : ______ minutes **Score :** ________ out of 36

1 6 - 1	2 2 - 1	3 5 - 7	4 1 - 6	5 8 - 8	6 3 - 8
7 10 - 2	8 8 - 7	9 9 - 3	10 8 - 5	11 8 - 2	12 1 - 3
13 6 - 8	14 5 - 5	15 6 - 3	16 6 - 3	17 5 - 1	18 0 - 3
19 8 - 7	20 10 - 9	21 3 - 6	22 7 - 0	23 8 - 7	24 10 - 6
25 4 - 3	26 6 - 2	27 1 - 1	28 0 - 9	29 10 - 1	30 10 - 0
31 0 - 8	32 5 - 7	33 8 - 10	34 8 - 3	35 0 - 5	36 8 - 5

Day 37

Subtraction digits 0-10

Time : ____ : ____ minutes **Score :** ________ out of 36

1. 5 − 5	2. 1 − 0	3. 5 − 7	4. 5 − 4	5. 5 − 9	6. 9 − 8
7. 6 − 9	8. 6 − 8	9. 2 − 4	10. 10 − 8	11. 10 − 7	12. 2 − 1
13. 1 − 0	14. 9 − 5	15. 7 − 7	16. 4 − 7	17. 9 − 0	18. 5 − 1
19. 10 − 2	20. 5 − 4	21. 1 − 2	22. 3 − 0	23. 7 − 3	24. 1 − 10
25. 9 − 9	26. 10 − 8	27. 9 − 10	28. 8 − 5	29. 0 − 1	30. 0 − 2
31. 7 − 10	32. 1 − 8	33. 0 − 10	34. 3 − 0	35. 2 − 2	36. 1 − 9

Day 38

Subtraction digits 0-10

Time : ____ : ____ minutes **Score :** ______ out of 36

1	2	3	4	5	6
9 - 0	9 - 7	2 - 2	10 - 0	8 - 7	0 - 4

7	8	9	10	11	12
3 - 1	1 - 2	5 - 3	0 - 8	4 - 7	8 - 0

13	14	15	16	17	18
7 - 0	0 - 3	2 - 2	0 - 10	2 - 8	6 - 2

19	20	21	22	23	24
5 - 10	2 - 8	9 - 0	7 - 0	0 - 2	10 - 4

25	26	27	28	29	30
7 - 9	2 - 6	6 - 5	4 - 8	4 - 3	8 - 1

31	32	33	34	35	36
9 - 6	0 - 0	0 - 8	8 - 6	3 - 5	2 - 5

Day 39
Subtraction digits 0-10

Time : ____ : ____ minutes **Score :** ______ out of 36

1	2	3	4	5	6
3 - 1	3 - 2	10 - 8	10 - 10	0 - 2	7 - 6

7	8	9	10	11	12
4 - 7	0 - 4	10 - 0	0 - 2	9 - 7	0 - 8

13	14	15	16	17	18
0 - 3	8 - 10	8 - 5	2 - 7	7 - 9	3 - 3

19	20	21	22	23	24
10 - 6	9 - 8	8 - 1	5 - 8	5 - 7	2 - 9

25	26	27	28	29	30
9 - 8	10 - 0	9 - 10	8 - 4	4 - 9	7 - 4

31	32	33	34	35	36
1 - 3	5 - 2	7 - 0	7 - 9	3 - 0	5 - 8

Day 40

Subtraction digits 0-10

Time : ____ : ____ minutes **Score :** ______ out of 36

1 2 - 5	2 9 - 5	3 1 - 5	4 0 - 0	5 5 - 0	6 3 - 4
7 7 - 7	8 0 - 3	9 10 - 7	10 0 - 7	11 5 - 8	12 5 - 9
13 3 - 3	14 4 - 0	15 9 - 7	16 7 - 5	17 9 - 1	18 5 - 4
19 2 - 7	20 1 - 6	21 4 - 6	22 7 - 9	23 7 - 8	24 3 - 9
25 9 - 5	26 2 - 5	27 5 - 8	28 9 - 9	29 5 - 1	30 9 - 1
31 10 - 10	32 8 - 5	33 10 - 4	34 6 - 9	35 7 - 8	36 5 - 3

Day 41

Subtraction digits 0-10

Time : ______:______ minutes **Score :** ________ out of 36

1 5 - 8	2 5 - 6	3 7 - 2	4 8 - 3	5 8 - 10	6 4 - 5
7 5 - 2	8 10 - 3	9 2 - 10	10 6 - 8	11 5 - 8	12 10 - 0
13 6 - 2	14 3 - 1	15 2 - 0	16 0 - 5	17 3 - 1	18 6 - 5
19 1 - 0	20 9 - 9	21 2 - 3	22 1 - 7	23 2 - 9	24 10 - 2
25 5 - 1	26 0 - 1	27 5 - 3	28 7 - 8	29 5 - 7	30 8 - 2
31 3 - 4	32 6 - 10	33 7 - 9	34 0 - 10	35 9 - 5	36 0 - 2

Day 42

Subtraction digits 0-10

Time : ____ : ____ minutes **Score :** ______ out of 36

1. 3 - 10	2. 3 - 4	3. 1 - 4	4. 8 - 7	5. 2 - 2	6. 0 - 8
7. 9 - 10	8. 9 - 3	9. 3 - 7	10. 9 - 3	11. 1 - 5	12. 1 - 9
13. 2 - 10	14. 3 - 4	15. 0 - 2	16. 3 - 3	17. 7 - 1	18. 3 - 8
19. 9 - 8	20. 8 - 6	21. 10 - 1	22. 9 - 0	23. 0 - 3	24. 5 - 7
25. 10 - 0	26. 10 - 8	27. 4 - 9	28. 6 - 6	29. 1 - 0	30. 10 - 7
31. 3 - 0	32. 2 - 6	33. 9 - 0	34. 7 - 7	35. 7 - 0	36. 3 - 5

Day 43

Subtraction digits 10-20

Time : ______ : ______ minutes **Score :** ________ out of 36

1	2	3	4	5	6
14 - 19	12 - 14	17 - 15	18 - 14	17 - 13	20 - 10

7	8	9	10	11	12
12 - 13	17 - 19	18 - 17	12 - 16	10 - 14	19 - 15

13	14	15	16	17	18
16 - 16	18 - 16	17 - 12	17 - 11	11 - 19	18 - 12

19	20	21	22	23	24
14 - 16	14 - 18	11 - 16	16 - 16	11 - 15	18 - 11

25	26	27	28	29	30
18 - 19	12 - 16	19 - 12	19 - 16	18 - 10	13 - 16

31	32	33	34	35	36
17 - 16	14 - 10	13 - 13	20 - 10	17 - 13	19 - 17

Day 44
Subtraction digits 10-20

Time : ______ : ______ minutes **Score :** ________ out of 36

1 11 - 15	2 10 - 15	3 12 - 14	4 16 - 14	5 11 - 16	6 13 - 11
7 18 - 12	8 10 - 17	9 20 - 14	10 15 - 14	11 17 - 13	12 13 - 18
13 15 - 12	14 19 - 16	15 10 - 12	16 11 - 10	17 16 - 17	18 16 - 15
19 10 - 15	20 11 - 18	21 12 - 14	22 16 - 14	23 17 - 10	24 18 - 11
25 20 - 20	26 19 - 18	27 17 - 15	28 18 - 16	29 20 - 17	30 20 - 10
31 20 - 12	32 17 - 16	33 15 - 20	34 19 - 20	35 13 - 19	36 16 - 10

Day 45

Subtraction digits 10-20

Time : ____:____ minutes **Score :** ________ out of 36

1	2	3	4	5	6
13 - 19	10 - 16	17 - 17	20 - 15	13 - 18	10 - 13

7	8	9	10	11	12
12 - 18	12 - 20	20 - 19	13 - 18	11 - 14	19 - 11

13	14	15	16	17	18
13 - 16	20 - 12	10 - 15	11 - 15	17 - 19	16 - 15

19	20	21	22	23	24
14 - 11	14 - 12	10 - 17	18 - 13	20 - 16	13 - 15

25	26	27	28	29	30
19 - 19	15 - 14	12 - 11	14 - 13	10 - 17	17 - 20

31	32	33	34	35	36
19 - 18	20 - 18	18 - 13	12 - 14	18 - 19	11 - 12

Day 46
Subtraction digits 10-20

Time : ______ : ______ minutes **Score :** ________ out of 36

1 20 - 15	2 13 - 10	3 12 - 12	4 12 - 16	5 10 - 17	6 13 - 16
7 16 - 18	8 12 - 17	9 17 - 10	10 15 - 15	11 11 - 17	12 16 - 17
13 15 - 20	14 14 - 13	15 17 - 13	16 15 - 18	17 15 - 17	18 19 - 13
19 13 - 14	20 19 - 14	21 15 - 14	22 13 - 14	23 16 - 18	24 17 - 10
25 13 - 13	26 16 - 16	27 10 - 13	28 17 - 11	29 15 - 16	30 19 - 12
31 18 - 18	32 19 - 10	33 20 - 12	34 10 - 12	35 19 - 13	36 14 - 14

Day 47
Subtraction digits 10-20

Time : ______:______ minutes **Score :** ________ out of 36

1 13 - 17	2 14 - 15	3 10 - 19	4 17 - 20	5 20 - 20	6 18 - 18
7 19 - 11	8 20 - 13	9 15 - 15	10 20 - 14	11 15 - 11	12 16 - 15
13 14 - 14	14 14 - 15	15 18 - 12	16 10 - 19	17 11 - 19	18 16 - 19
19 11 - 20	20 10 - 16	21 17 - 14	22 10 - 15	23 13 - 14	24 20 - 16
25 18 - 20	26 20 - 10	27 11 - 15	28 15 - 13	29 10 - 16	30 17 - 20
31 20 - 10	32 11 - 12	33 16 - 12	34 17 - 15	35 19 - 13	36 16 - 18

Day 48

Subtraction digits 10-20

Time : ____ : ____ minutes **Score :** ______ out of 36

1	2	3	4	5	6
10 - 15 =	19 - 18 =	16 - 14 =	14 - 14 =	18 - 18 =	14 - 12 =

7	8	9	10	11	12
16 - 11 =	15 - 15 =	16 - 12 =	20 - 12 =	12 - 11 =	13 - 15 =

13	14	15	16	17	18
15 - 19 =	16 - 20 =	18 - 10 =	13 - 13 =	11 - 20 =	18 - 16 =

19	20	21	22	23	24
20 - 14 =	14 - 10 =	13 - 10 =	14 - 13 =	12 - 18 =	12 - 13 =

25	26	27	28	29	30
13 - 19 =	18 - 17 =	19 - 13 =	11 - 15 =	15 - 20 =	18 - 15 =

31	32	33	34	35	36
11 - 13 =	14 - 20 =	11 - 15 =	18 - 12 =	16 - 16 =	13 - 17 =

Day 49
Subtraction digits 10-20

Time : ____ : ____ minutes **Score :** ______ out of 36

1	2	3	4	5	6
13 - 19	14 - 14	20 - 11	13 - 11	18 - 12	18 - 18

7	8	9	10	11	12
20 - 10	14 - 16	13 - 19	19 - 17	16 - 20	19 - 10

13	14	15	16	17	18
16 - 18	10 - 11	17 - 15	11 - 13	13 - 10	16 - 19

19	20	21	22	23	24
10 - 10	18 - 11	17 - 16	20 - 15	12 - 11	13 - 13

25	26	27	28	29	30
19 - 20	20 - 13	15 - 18	20 - 20	20 - 12	10 - 18

31	32	33	34	35	36
18 - 17	10 - 15	13 - 12	16 - 13	17 - 19	14 - 12

Day 50
Subtraction digits 10-20

Time : ____:____ minutes **Score :** ________ out of 36

1. 13 - 20	2. 10 - 18	3. 14 - 10	4. 10 - 12	5. 14 - 18	6. 15 - 13
7. 11 - 19	8. 11 - 20	9. 19 - 12	10. 18 - 14	11. 15 - 17	12. 14 - 17
13. 18 - 15	14. 14 - 12	15. 13 - 19	16. 13 - 12	17. 20 - 14	18. 16 - 11
19. 18 - 15	20. 18 - 18	21. 17 - 14	22. 20 - 10	23. 14 - 20	24. 16 - 18
25. 15 - 15	26. 12 - 14	27. 10 - 14	28. 19 - 14	29. 19 - 17	30. 11 - 11
31. 13 - 20	32. 17 - 13	33. 20 - 20	34. 12 - 11	35. 17 - 17	36. 15 - 20

Day 51
Subtraction digits 10-20

Time : ______ : ______ minutes **Score :** ________ out of 36

1	2	3	4	5	6
16 - 10	13 - 19	12 - 15	17 - 16	14 - 14	11 - 19

7	8	9	10	11	12
17 - 17	15 - 20	12 - 11	13 - 12	13 - 14	15 - 15

13	14	15	16	17	18
16 - 13	15 - 20	18 - 13	15 - 13	15 - 10	20 - 19

19	20	21	22	23	24
17 - 11	20 - 13	19 - 19	14 - 12	15 - 19	14 - 10

25	26	27	28	29	30
20 - 19	10 - 16	13 - 19	13 - 13	18 - 17	13 - 17

31	32	33	34	35	36
16 - 19	10 - 15	18 - 11	20 - 17	13 - 14	15 - 11

Day 52

Subtraction digits 10-20

Time : ______ : ______ minutes **Score :** ________ out of 36

1	2	3	4	5	6
18 - 18	10 - 20	10 - 14	12 - 16	20 - 13	18 - 16
7	8	9	10	11	12
20 - 14	12 - 10	17 - 15	11 - 15	16 - 14	20 - 11
13	14	15	16	17	18
14 - 11	15 - 17	13 - 18	13 - 12	11 - 11	16 - 17
19	20	21	22	23	24
10 - 14	14 - 20	10 - 11	12 - 14	10 - 10	17 - 20
25	26	27	28	29	30
18 - 20	12 - 11	17 - 19	11 - 16	12 - 17	20 - 18
31	32	33	34	35	36
19 - 19	12 - 13	14 - 11	14 - 15	19 - 12	13 - 11

Day 53
Subtraction digits 10-20

Time : ______ : ______ minutes **Score :** ________ out of 36

1	2	3	4	5	6
12 - 14	14 - 11	11 - 17	12 - 15	11 - 17	10 - 15
7	8	9	10	11	12
19 - 18	14 - 19	12 - 17	10 - 13	20 - 17	20 - 10
13	14	15	16	17	18
13 - 12	20 - 13	20 - 16	11 - 15	12 - 20	13 - 16
19	20	21	22	23	24
19 - 20	17 - 13	11 - 11	17 - 11	19 - 13	16 - 18
25	26	27	28	29	30
14 - 16	13 - 13	15 - 19	20 - 12	11 - 11	15 - 19
31	32	33	34	35	36
10 - 11	13 - 12	12 - 19	14 - 16	13 - 18	17 - 18

Day 54

Subtraction digits 10-20

Time : ______ : ______ minutes **Score :** ________ out of 36

1	2	3	4	5	6
18 - 18	16 - 13	14 - 19	10 - 13	10 - 17	16 - 18

7	8	9	10	11	12
15 - 17	13 - 15	20 - 20	12 - 13	17 - 16	18 - 11

13	14	15	16	17	18
15 - 16	12 - 16	13 - 16	15 - 13	12 - 19	20 - 13

19	20	21	22	23	24
16 - 14	10 - 16	20 - 20	13 - 20	12 - 14	15 - 10

25	26	27	28	29	30
13 - 13	20 - 20	20 - 18	10 - 18	10 - 16	18 - 20

31	32	33	34	35	36
13 - 16	18 - 13	18 - 20	15 - 18	15 - 17	14 - 12

Day 55

Subtraction digits 10-20

Time : ____:____ minutes **Score :** ______ out of 36

1. 16 - 14	2. 15 - 17	3. 14 - 13	4. 14 - 16	5. 12 - 17	6. 10 - 10
7. 15 - 20	8. 13 - 16	9. 15 - 16	10. 19 - 10	11. 19 - 10	12. 19 - 16
13. 19 - 14	14. 10 - 10	15. 18 - 16	16. 14 - 18	17. 13 - 13	18. 14 - 15
19. 14 - 18	20. 12 - 10	21. 13 - 17	22. 16 - 18	23. 14 - 19	24. 12 - 15
25. 16 - 12	26. 12 - 16	27. 20 - 14	28. 13 - 19	29. 16 - 17	30. 15 - 14
31. 15 - 12	32. 15 - 13	33. 12 - 13	34. 14 - 12	35. 16 - 19	36. 17 - 12

Day 56

Subtraction digits 10-20

Time : ______ : ______ minutes **Score :** ________ out of 36

1 10 - 16	2 15 - 12	3 20 - 20	4 19 - 12	5 14 - 19	6 14 - 10
7 13 - 20	8 16 - 17	9 17 - 13	10 15 - 20	11 16 - 17	12 20 - 14
13 20 - 18	14 20 - 17	15 14 - 18	16 20 - 12	17 12 - 13	18 15 - 20
19 15 - 15	20 10 - 10	21 13 - 16	22 14 - 16	23 13 - 13	24 14 - 13
25 19 - 10	26 19 - 17	27 16 - 12	28 12 - 19	29 15 - 20	30 10 - 11
31 13 - 17	32 13 - 18	33 11 - 20	34 16 - 17	35 12 - 12	36 17 - 19

Day 57

Subtraction digits 0-20

Time : ______ : ______ minutes **Score :** ________ out of 36

1 5 - 1	2 1 - 16	3 16 - 10	4 5 - 13	5 15 - 12	6 3 - 4
7 17 - 2	8 5 - 3	9 0 - 6	10 3 - 1	11 19 - 10	12 1 - 11
13 1 - 19	14 10 - 17	15 14 - 14	16 12 - 14	17 6 - 18	18 18 - 18
19 1 - 9	20 20 - 18	21 14 - 15	22 10 - 9	23 10 - 9	24 15 - 14
25 2 - 20	26 2 - 15	27 18 - 14	28 13 - 0	29 17 - 20	30 19 - 7
31 3 - 3	32 3 - 8	33 8 - 2	34 8 - 9	35 7 - 1	36 5 - 0

Day 58

Subtraction digits 0-20

Time : ______ : ______ minutes **Score :** ________ out of 36

1	2	3	4	5	6
1 - 6	8 - 9	3 - 2	7 - 20	12 - 12	15 - 13

7	8	9	10	11	12
13 - 7	2 - 18	4 - 17	10 - 10	15 - 14	16 - 16

13	14	15	16	17	18
4 - 11	5 - 9	3 - 1	7 - 8	8 - 12	6 - 3

19	20	21	22	23	24
15 - 6	13 - 6	8 - 7	3 - 3	6 - 11	16 - 4

25	26	27	28	29	30
19 - 0	16 - 10	9 - 18	3 - 12	3 - 16	5 - 11

31	32	33	34	35	36
5 - 3	1 - 0	14 - 19	3 - 4	8 - 12	18 - 6

Day 59
Subtraction digits 0-20

Time : ______ : ______ minutes **Score :** ________ out of 36

1	2	3	4	5	6
15 - 16 =	16 - 15 =	19 - 6 =	18 - 2 =	18 - 17 =	7 - 17 =

7	8	9	10	11	12
15 - 7 =	15 - 9 =	3 - 11 =	10 - 11 =	6 - 11 =	6 - 10 =

13	14	15	16	17	18
7 - 1 =	17 - 3 =	0 - 14 =	0 - 9 =	4 - 12 =	7 - 3 =

19	20	21	22	23	24
6 - 18 =	13 - 19 =	2 - 14 =	0 - 5 =	9 - 18 =	8 - 15 =

25	26	27	28	29	30
2 - 3 =	15 - 11 =	16 - 2 =	18 - 13 =	18 - 13 =	19 - 15 =

31	32	33	34	35	36
13 - 20 =	14 - 0 =	1 - 3 =	0 - 11 =	13 - 9 =	20 - 5 =

Day 60

Subtraction digits 0-20

Time : ____ : ____ minutes **Score :** ________ out of 36

1	2	3	4	5	6
9 - 12	8 - 20	10 - 14	5 - 12	4 - 11	6 - 11

7	8	9	10	11	12
18 - 18	7 - 11	7 - 8	10 - 18	17 - 9	11 - 16

13	14	15	16	17	18
9 - 2	8 - 17	20 - 16	7 - 1	12 - 4	5 - 11

19	20	21	22	23	24
3 - 18	17 - 2	3 - 6	15 - 5	12 - 15	17 - 11

25	26	27	28	29	30
1 - 16	13 - 8	16 - 13	15 - 1	3 - 18	16 - 11

31	32	33	34	35	36
3 - 4	5 - 7	8 - 0	15 - 10	13 - 10	3 - 20

Day 61

Subtraction digits 0-20

Time : ____:____ minutes **Score :** ________ out of 36

1	2	3	4	5	6
19 - 0	8 - 3	2 - 9	13 - 17	4 - 3	6 - 13

7	8	9	10	11	12
16 - 13	7 - 15	14 - 11	2 - 7	18 - 16	1 - 10

13	14	15	16	17	18
10 - 18	20 - 9	20 - 1	3 - 14	5 - 16	0 - 18

19	20	21	22	23	24
1 - 17	19 - 18	2 - 9	0 - 20	13 - 7	14 - 14

25	26	27	28	29	30
12 - 12	1 - 14	14 - 13	11 - 7	0 - 15	14 - 2

31	32	33	34	35	36
18 - 16	13 - 15	20 - 15	4 - 12	2 - 12	14 - 7

Day 62

Subtraction digits 0-20

Time : ______:______ minutes **Score :** ________ out of 36

1 11 - 16	2 6 - 14	3 10 - 2	4 20 - 7	5 19 - 13	6 14 - 10
7 8 - 20	8 3 - 6	9 2 - 7	10 4 - 5	11 17 - 11	12 11 - 19
13 13 - 8	14 12 - 3	15 20 - 19	16 12 - 11	17 13 - 2	18 0 - 7
19 3 - 9	20 12 - 12	21 16 - 5	22 18 - 0	23 15 - 5	24 20 - 8
25 12 - 12	26 11 - 17	27 18 - 6	28 18 - 1	29 8 - 5	30 13 - 7
31 15 - 1	32 3 - 3	33 5 - 13	34 4 - 18	35 11 - 8	36 19 - 5

Day 63

Subtraction digits 0-20

Time : ____ : ____ minutes **Score :** ________ out of 36

1	2	3	4	5	6
19 - 11	2 - 20	18 - 15	13 - 15	17 - 16	0 - 16
7	8	9	10	11	12
2 - 4	18 - 8	6 - 5	17 - 12	18 - 12	16 - 20
13	14	15	16	17	18
17 - 13	8 - 7	11 - 2	20 - 3	13 - 10	5 - 13
19	20	21	22	23	24
4 - 2	13 - 15	16 - 6	1 - 2	18 - 2	18 - 5
25	26	27	28	29	30
12 - 19	9 - 13	15 - 10	19 - 3	9 - 8	6 - 8
31	32	33	34	35	36
15 - 10	6 - 12	4 - 17	7 - 4	8 - 17	0 - 17

Day 64

Subtraction digits 0-20

Time : ______ : ______ minutes **Score :** ________ out of 36

1. 13 - 2 =	2. 17 - 20 =	3. 19 - 10 =	4. 11 - 4 =	5. 5 - 17 =	6. 1 - 14 =
7. 13 - 5 =	8. 1 - 10 =	9. 3 - 5 =	10. 17 - 6 =	11. 13 - 5 =	12. 7 - 18 =
13. 16 - 12 =	14. 12 - 16 =	15. 5 - 11 =	16. 11 - 13 =	17. 17 - 19 =	18. 12 - 7 =
19. 18 - 6 =	20. 9 - 13 =	21. 6 - 1 =	22. 10 - 16 =	23. 17 - 8 =	24. 2 - 17 =
25. 20 - 10 =	26. 15 - 0 =	27. 20 - 1 =	28. 8 - 4 =	29. 17 - 3 =	30. 6 - 9 =
31. 11 - 16 =	32. 10 - 16 =	33. 8 - 15 =	34. 7 - 1 =	35. 14 - 8 =	36. 8 - 2 =

Day 65

Subtraction digits 0-20

Time : ______ : ______ minutes **Score :** ________ out of 36

1. 0 - 1	2. 10 - 15	3. 2 - 18	4. 20 - 5	5. 9 - 0	6. 10 - 18
7. 4 - 18	8. 17 - 19	9. 19 - 13	10. 2 - 6	11. 5 - 13	12. 4 - 17
13. 10 - 1	14. 14 - 5	15. 9 - 6	16. 16 - 18	17. 2 - 18	18. 19 - 16
19. 20 - 5	20. 7 - 20	21. 4 - 13	22. 16 - 11	23. 0 - 16	24. 0 - 3
25. 17 - 6	26. 6 - 15	27. 8 - 20	28. 19 - 20	29. 1 - 11	30. 6 - 5
31. 4 - 4	32. 17 - 18	33. 4 - 13	34. 14 - 7	35. 3 - 1	36. 6 - 15

Day 66

Subtraction digits 0-20

Time : ____:____ minutes **Score :** ________ out of 36

1. 20 - 4 =	2. 4 - 6 =	3. 12 - 20 =	4. 8 - 17 =	5. 7 - 11 =	6. 13 - 1 =
7. 15 - 1 =	8. 19 - 3 =	9. 16 - 17 =	10. 19 - 11 =	11. 16 - 11 =	12. 20 - 19 =
13. 6 - 7 =	14. 19 - 10 =	15. 5 - 11 =	16. 19 - 14 =	17. 18 - 10 =	18. 5 - 15 =
19. 15 - 3 =	20. 19 - 20 =	21. 11 - 15 =	22. 18 - 12 =	23. 20 - 12 =	24. 9 - 10 =
25. 14 - 20 =	26. 10 - 17 =	27. 19 - 2 =	28. 2 - 19 =	29. 13 - 0 =	30. 18 - 5 =
31. 0 - 20 =	32. 4 - 3 =	33. 0 - 10 =	34. 2 - 12 =	35. 17 - 12 =	36. 10 - 3 =

Day 67
Subtraction digits 0-20

Time : ______ : ______ minutes **Score :** ________ out of 36

1. 5 - 11 =	2. 19 - 3 =	3. 13 - 5 =	4. 2 - 2 =	5. 18 - 4 =	6. 13 - 5 =
7. 12 - 2 =	8. 18 - 18 =	9. 6 - 16 =	10. 7 - 4 =	11. 8 - 2 =	12. 8 - 2 =
13. 3 - 16 =	14. 10 - 14 =	15. 19 - 1 =	16. 18 - 15 =	17. 16 - 16 =	18. 19 - 1 =
19. 15 - 11 =	20. 14 - 1 =	21. 10 - 2 =	22. 16 - 7 =	23. 9 - 16 =	24. 6 - 7 =
25. 19 - 7 =	26. 4 - 20 =	27. 16 - 4 =	28. 13 - 14 =	29. 11 - 4 =	30. 12 - 6 =
31. 19 - 13 =	32. 11 - 2 =	33. 2 - 6 =	34. 10 - 7 =	35. 6 - 8 =	36. 20 - 20 =

Day 68
Subtraction digits 0-20

Time : ____ : ____ minutes **Score :** ______ out of 36

1	2	3	4	5	6
7 - 6	14 - 7	19 - 17	5 - 14	3 - 6	20 - 14

7	8	9	10	11	12
19 - 4	18 - 18	3 - 8	3 - 11	10 - 17	12 - 15

13	14	15	16	17	18
15 - 4	3 - 17	9 - 13	13 - 12	9 - 17	8 - 5

19	20	21	22	23	24
18 - 8	10 - 6	20 - 10	12 - 19	10 - 11	19 - 11

25	26	27	28	29	30
13 - 2	13 - 20	16 - 6	7 - 8	5 - 13	2 - 17

31	32	33	34	35	36
15 - 10	5 - 9	13 - 20	7 - 12	3 - 11	5 - 16

Day 69

Subtraction digits 0-20

Time : ______ : ______ minutes **Score :** ________ out of 36

1	2	3	4	5	6
7 - 1 =	19 - 12 =	6 - 0 =	2 - 9 =	6 - 19 =	15 - 5 =

7	8	9	10	11	12
11 - 17 =	9 - 17 =	14 - 11 =	16 - 3 =	9 - 8 =	0 - 3 =

13	14	15	16	17	18
18 - 7 =	15 - 7 =	16 - 6 =	5 - 1 =	16 - 20 =	5 - 12 =

19	20	21	22	23	24
11 - 4 =	9 - 2 =	3 - 13 =	17 - 1 =	14 - 3 =	10 - 7 =

25	26	27	28	29	30
20 - 4 =	19 - 16 =	1 - 6 =	1 - 17 =	1 - 3 =	13 - 3 =

31	32	33	34	35	36
6 - 2 =	0 - 5 =	6 - 6 =	17 - 20 =	14 - 2 =	11 - 1 =

Day 70
Subtraction digits 0-20

Time : ____ : ____ minutes **Score :** ______ out of 36

1	2	3	4	5	6
17 - 19	20 - 4	17 - 1	12 - 19	2 - 6	3 - 8
7	8	9	10	11	12
11 - 13	12 - 5	4 - 9	5 - 3	16 - 10	18 - 18
13	14	15	16	17	18
1 - 20	2 - 9	14 - 8	6 - 4	4 - 16	8 - 2
19	20	21	22	23	24
20 - 13	16 - 1	3 - 6	2 - 10	13 - 10	12 - 19
25	26	27	28	29	30
18 - 5	2 - 20	17 - 5	19 - 6	2 - 5	5 - 2
31	32	33	34	35	36
18 - 11	12 - 0	13 - 10	11 - 8	12 - 12	11 - 3

SOLUTION

Day 1	Day 2	Day 3	Day 4	Day 5	Day 6	Day 7
1 = 10	1 = 4	1 = 6	1 = 5	1 = 5	1 = 5	1 = 9
2 = 0	2 = 1	2 = 5	2 = 3	2 = 1	2 = 3	2 = 5
3 = 9	3 = 1	3 = 2	3 = 6	3 = 6	3 = 0	3 = 2
4 = 2	4 = 4	4 = 0	4 = 5	4 = 5	4 = 0	4 = 7
5 = 6	5 = 4	5 = 1	5 = 2	5 = 2	5 = 8	5 = 5
6 = 9	6 = 6	6 = 2	6 = 7	6 = 2	6 = 2	6 = 7
7 = 6	7 = 6	7 = 3	7 = 7	7 = 3	7 = 5	7 = 2
8 = 9	8 = 5	8 = 4	8 = 9	8 = 4	8 = 7	8 = 3
9 = 4	9 = 3	9 = 5	9 = 2	9 = 4	9 = 8	9 = 4
10 = 3	10 = 2	10 = 2	10 = 8	10 = 5	10 = 3	10 = 5
11 = 2	11 = 4	11 = 1	11 = 5	11 = 3	11 = 9	11 = 6
12 = 5	12 = 2	12 = 7	12 = 3	12 = 8	12 = 6	12 = 8
13 = 5	13 = 6	13 = 5	13 = 3	13 = 1	13 = 2	13 = 7
14 = 5	14 = 10	14 = 3	14 = 2	14 = 8	14 = 5	14 = 7
15 = 5	15 = 5	15 = 3	15 = 10	15 = 4	15 = 7	15 = 8
16 = 4	16 = 5	16 = 8	16 = 9	16 = 5	16 = 3	16 = 8
17 = 5	17 = 1	17 = 6	17 = 4	17 = 5	17 = 4	17 = 3
18 = 7	18 = 6	18 = 4	18 = 1	18 = 6	18 = 6	18 = 9
19 = 5	19 = 2	19 = 5	19 = 5	19 = 3	19 = 0	19 = 9
20 = 5	20 = 5	20 = 5	20 = 8	20 = 1	20 = 6	20 = 5
21 = 1	21 = 5	21 = 1	21 = 6	21 = 5	21 = 5	21 = 4
22 = 4	22 = 3	22 = 6	22 = 3	22 = 2	22 = 1	22 = 2
23 = 0	23 = 4	23 = 10	23 = 6	23 = 9	23 = 7	23 = 5
24 = 6	24 = 0	24 = 0	24 = 2	24 = 2	24 = 1	24 = 7
25 = 2	25 = 5	25 = 7	25 = 2	25 = 8	25 = 6	25 = 4
26 = 6	26 = 7	26 = 2	26 = 7	26 = 3	26 = 7	26 = 4
27 = 4	27 = 8	27 = 3	27 = 8	27 = 5	27 = 4	27 = 6
28 = 5	28 = 8	28 = 8	28 = 4	28 = 0	28 = 3	28 = 10
29 = 4	29 = 3	29 = 5	29 = 2	29 = 6	29 = 6	29 = 3
30 = 4	30 = 3	30 = 6	30 = 3	30 = 5	30 = 0	30 = 5
31 = 5	31 = 0	31 = 5	31 = 5	31 = 6	31 = 3	31 = 3
32 = 7	32 = 4	32 = 5	32 = 9	32 = 5	32 = 9	32 = 9
33 = 7	33 = 4	33 = 2	33 = 7	33 = 2	33 = 3	33 = 5
34 = 3	34 = 5	34 = 3	34 = 10	34 = 9	34 = 5	34 = 5
35 = 9	35 = 4	35 = 8	35 = 4	35 = 8	35 = 5	35 = 6
36 = 4	36 = 7	36 = 4	36 = 3	36 = 8	36 = 3	36 = 4

Day 8	Day 9	Day 10	Day 11	Day 12	Day 13	Day 14
1 = 5	1 = 5	1 = 4	1 = 14	1 = 5	1 = 6	1 = 10
2 = 11	2 = 3	2 = 13	2 = 4	2 = 3	2 = 2	2 = 10
3 = 2	3 = 11	3 = 8	3 = 7	3 = 8	3 = 5	3 = 5
4 = 7	4 = 10	4 = 11	4 = 7	4 = 2	4 = 6	4 = 7
5 = 7	5 = 5	5 = 6	5 = 7	5 = 5	5 = 0	5 = 13
6 = 4	6 = 5	6 = 11	6 = 3	6 = 10	6 = 6	6 = 3
7 = 11	7 = 9	7 = 8	7 = 6	7 = 3	7 = 12	7 = 8
8 = 11	8 = 8	8 = 6	8 = 11	8 = 5	8 = 6	8 = 7
9 = 9	9 = 11	9 = 9	9 = 7	9 = 11	9 = 11	9 = 8
10 = 9	10 = 2	10 = 11	10 = 6	10 = 2	10 = 2	10 = 0
11 = 3	11 = 8	11 = 7	11 = 9	11 = 7	11 = 5	11 = 7
12 = 10	12 = 7	12 = 4	12 = 8	12 = 8	12 = 8	12 = 8
13 = 13	13 = 5	13 = 9	13 = 6	13 = 9	13 = 12	13 = 4
14 = 6	14 = 3	14 = 4	14 = 10	14 = 8	14 = 3	14 = 5
15 = 7	15 = 10	15 = 9	15 = 1	15 = 5	15 = 11	15 = 7
16 = 4	16 = 6	16 = 1	16 = 2	16 = 8	16 = 5	16 = 4
17 = 7	17 = 6	17 = 10	17 = 9	17 = 12	17 = 12	17 = 11
18 = 9	18 = 7	18 = 9	18 = 7	18 = 10	18 = 3	18 = 10
19 = 8	19 = 12	19 = 3	19 = 5	19 = 4	19 = 0	19 = 4
20 = 9	20 = 7	20 = 12	20 = 5	20 = 7	20 = 0	20 = 0
21 = 5	21 = 8	21 = 9	21 = 6	21 = 3	21 = 7	21 = 4
22 = 5	22 = 2	22 = 6	22 = 11	22 = 13	22 = 4	22 = 6
23 = 12	23 = 6	23 = 3	23 = 4	23 = 3	23 = 6	23 = 10
24 = 9	24 = 10	24 = 2	24 = 8	24 = 10	24 = 8	24 = 4
25 = 5	25 = 5	25 = 10	25 = 7	25 = 5	25 = 7	25 = 8
26 = 3	26 = 10	26 = 6	26 = 4	26 = 0	26 = 8	26 = 3
27 = 9	27 = 7	27 = 11	27 = 2	27 = 5	27 = 11	27 = 6
28 = 7	28 = 14	28 = 7	28 = 6	28 = 12	28 = 3	28 = 12
29 = 4	29 = 11	29 = 5	29 = 8	29 = 0	29 = 9	29 = 5
30 = 8	30 = 2	30 = 12	30 = 13	30 = 7	30 = 8	30 = 9
31 = 10	31 = 3	31 = 7	31 = 4	31 = 3	31 = 6	31 = 6
32 = 6	32 = 5	32 = 3	32 = 13	32 = 6	32 = 5	32 = 13
33 = 7	33 = 9	33 = 3	33 = 7	33 = 5	33 = 5	33 = 6
34 = 4	34 = 12	34 = 11	34 = 12	34 = 8	34 = 12	34 = 7
35 = 2	35 = 11	35 = 1	35 = 8	35 = 11	35 = 7	35 = 2
36 = 9	36 = 6	36 = 5	36 = 7	36 = 9	36 = 8	36 = 7

Day 15	Day 16	Day 17	Day 18	Day 19	Day 20	Day 21
1 = 10	1 = 7	1 = 10	1 = 5	1 = 18	1 = 10	1 = 13
2 = 9	2 = 6	2 = 15	2 = 8	2 = 7	2 = 15	2 = 8
3 = 10	3 = 3	3 = 6	3 = 17	3 = 6	3 = 4	3 = 9
4 = 14	4 = 5	4 = 14	4 = 4	4 = 2	4 = 7	4 = 10
5 = 6	5 = 9	5 = 8	5 = 13	5 = 2	5 = 14	5 = 2
6 = 10	6 = 18	6 = 4	6 = 16	6 = 6	6 = 6	6 = 11
7 = 4	7 = 10	7 = 12	7 = 16	7 = 13	7 = 13	7 = 11
8 = 11	8 = 5	8 = 7	8 = 17	8 = 7	8 = 6	8 = 9
9 = 6	9 = 7	9 = 14	9 = 7	9 = 13	9 = 13	9 = 9
10 = 15	10 = 12	10 = 4	10 = 9	10 = 10	10 = 12	10 = 10
11 = 15	11 = 9	11 = 10	11 = 15	11 = 9	11 = 10	11 = 6
12 = 9	12 = 13	12 = 8	12 = 2	12 = 14	12 = 11	12 = 14
13 = 4	13 = 16	13 = 14	13 = 17	13 = 12	13 = 19	13 = 11
14 = 18	14 = 7	14 = 7	14 = 11	14 = 3	14 = 2	14 = 16
15 = 13	15 = 2	15 = 8	15 = 12	15 = 7	15 = 6	15 = 11
16 = 10	16 = 8	16 = 0	16 = 9	16 = 6	16 = 15	16 = 7
17 = 1	17 = 4	17 = 10	17 = 8	17 = 15	17 = 1	17 = 8
18 = 10	18 = 7	18 = 14	18 = 8	18 = 10	18 = 15	18 = 8
19 = 19	19 = 16	19 = 11	19 = 2	19 = 11	19 = 6	19 = 4
20 = 2	20 = 8	20 = 9	20 = 12	20 = 8	20 = 10	20 = 16
21 = 12	21 = 8	21 = 9	21 = 16	21 = 16	21 = 1	21 = 9
22 = 15	22 = 7	22 = 16	22 = 8	22 = 18	22 = 8	22 = 10
23 = 5	23 = 10	23 = 12	23 = 10	23 = 3	23 = 16	23 = 12
24 = 7	24 = 7	24 = 5	24 = 5	24 = 2	24 = 13	24 = 14
25 = 12	25 = 1	25 = 10	25 = 7	25 = 19	25 = 14	25 = 7
26 = 17	26 = 3	26 = 3	26 = 3	26 = 12	26 = 13	26 = 14
27 = 11	27 = 8	27 = 4	27 = 4	27 = 10	27 = 7	27 = 11
28 = 15	28 = 15	28 = 6	28 = 8	28 = 9	28 = 15	28 = 13
29 = 10	29 = 11	29 = 13	29 = 8	29 = 9	29 = 16	29 = 9
30 = 17	30 = 11	30 = 11	30 = 4	30 = 10	30 = 5	30 = 8
31 = 1	31 = 7	31 = 15	31 = 8	31 = 1	31 = 13	31 = 16
32 = 4	32 = 17	32 = 14	32 = 20	32 = 10	32 = 14	32 = 11
33 = 9	33 = 10	33 = 10	33 = 7	33 = 20	33 = 9	33 = 6
34 = 10	34 = 10	34 = 9	34 = 13	34 = 8	34 = 12	34 = 4
35 = 15	35 = 14	35 = 8	35 = 5	35 = 13	35 = 3	35 = 7
36 = 8	36 = 7	36 = 3	36 = 18	36 = 14	36 = 10	36 = 11

Day 22	Day 23	Day 24	Day 25	Day 26	Day 27	Day 28
1 = 13	1 = 12	1 = 1	1 = 15	1 = 17	1 = 15	1 = 11
2 = 3	2 = 10	2 = 14	2 = 3	2 = 12	2 = 8	2 = 18
3 = 13	3 = 6	3 = 8	3 = 7	3 = 9	3 = 8	3 = 8
4 = 8	4 = 12	4 = 9	4 = 8	4 = 5	4 = 15	4 = 13
5 = 8	5 = 8	5 = 11	5 = 5	5 = 13	5 = 10	5 = 5
6 = 11	6 = 16	6 = 1	6 = 12	6 = 7	6 = 13	6 = 7
7 = 11	7 = 14	7 = 12	7 = 16	7 = 11	7 = 12	7 = 12
8 = 16	8 = 4	8 = 2	8 = 9	8 = 12	8 = 18	8 = 12
9 = 13	9 = 6	9 = 17	9 = 7	9 = 7	9 = 18	9 = 1
10 = 10	10 = 8	10 = 17	10 = 17	10 = 12	10 = 9	10 = 3
11 = 16	11 = 13	11 = 12	11 = 13	11 = 11	11 = 0	11 = 7
12 = 13	12 = 16	12 = 15	12 = 8	12 = 11	12 = 9	12 = 5
13 = 6	13 = 13	13 = 7	13 = 15	13 = 17	13 = 11	13 = 19
14 = 9	14 = 10	14 = 10	14 = 14	14 = 7	14 = 10	14 = 8
15 = 15	15 = 5	15 = 18	15 = 11	15 = 6	15 = 16	15 = 11
16 = 12	16 = 12	16 = 2	16 = 14	16 = 9	16 = 9	16 = 7
17 = 8	17 = 11	17 = 14	17 = 7	17 = 12	17 = 12	17 = 9
18 = 10	18 = 11	18 = 11	18 = 16	18 = 3	18 = 14	18 = 15
19 = 7	19 = 11	19 = 8	19 = 17	19 = 15	19 = 11	19 = 7
20 = 4	20 = 13	20 = 12	20 = 14	20 = 9	20 = 11	20 = 7
21 = 10	21 = 13	21 = 3	21 = 11	21 = 17	21 = 10	21 = 8
22 = 5	22 = 9	22 = 2	22 = 2	22 = 12	22 = 17	22 = 9
23 = 9	23 = 4	23 = 9	23 = 4	23 = 5	23 = 4	23 = 3
24 = 2	24 = 2	24 = 5	24 = 5	24 = 7	24 = 13	24 = 5
25 = 9	25 = 14	25 = 16	25 = 5	25 = 10	25 = 9	25 = 12
26 = 12	26 = 3	26 = 12	26 = 6	26 = 0	26 = 18	26 = 10
27 = 5	27 = 10	27 = 17	27 = 2	27 = 5	27 = 16	27 = 5
28 = 12	28 = 15	28 = 13	28 = 11	28 = 12	28 = 16	28 = 9
29 = 14	29 = 17	29 = 15	29 = 8	29 = 13	29 = 10	29 = 2
30 = 12	30 = 9	30 = 3	30 = 8	30 = 9	30 = 11	30 = 12
31 = 9	31 = 7	31 = 8	31 = 5	31 = 3	31 = 6	31 = 8
32 = 5	32 = 5	32 = 4	32 = 6	32 = 7	32 = 13	32 = 4
33 = 5	33 = 15	33 = 14	33 = 15	33 = 8	33 = 4	33 = 6
34 = 14	34 = 7	34 = 10	34 = 17	34 = 17	34 = 11	34 = 13
35 = 18	35 = 9	35 = 8	35 = 16	35 = 8	35 = 7	35 = 4
36 = 7	36 = 7	36 = 10	36 = 18	36 = 7	36 = 12	36 = 8

Day		29	Day		30	Day		31	Day		32	Day		33	Day		34	Day		35
1	=	1	1	=	8	1	=	6	1	=	15	1	=	13	1	=	14	1	=	5
2	=	10	2	=	8	2	=	6	2	=	7	2	=	12	2	=	6	2	=	14
3	=	9	3	=	8	3	=	10	3	=	15	3	=	4	3	=	7	3	=	6
4	=	12	4	=	7	4	=	10	4	=	19	4	=	6	4	=	9	4	=	11
5	=	11	5	=	10	5	=	14	5	=	14	5	=	10	5	=	6	5	=	8
6	=	6	6	=	10	6	=	6	6	=	11	6	=	9	6	=	10	6	=	7
7	=	16	7	=	9	7	=	12	7	=	16	7	=	7	7	=	11	7	=	6
8	=	10	8	=	5	8	=	11	8	=	12	8	=	10	8	=	8	8	=	4
9	=	14	9	=	8	9	=	16	9	=	1	9	=	15	9	=	15	9	=	6
10	=	10	10	=	11	10	=	17	10	=	7	10	=	9	10	=	3	10	=	8
11	=	3	11	=	14	11	=	17	11	=	13	11	=	16	11	=	13	11	=	12
12	=	9	12	=	12	12	=	6	12	=	5	12	=	6	12	=	6	12	=	13
13	=	7	13	=	5	13	=	9	13	=	12	13	=	3	13	=	12	13	=	13
14	=	9	14	=	5	14	=	10	14	=	10	14	=	16	14	=	15	14	=	5
15	=	11	15	=	13	15	=	8	15	=	7	15	=	10	15	=	17	15	=	8
16	=	9	16	=	11	16	=	10	16	=	9	16	=	7	16	=	11	16	=	9
17	=	3	17	=	8	17	=	11	17	=	11	17	=	11	17	=	9	17	=	16
18	=	7	18	=	6	18	=	10	18	=	10	18	=	10	18	=	8	18	=	11
19	=	10	19	=	14	19	=	6	19	=	13	19	=	4	19	=	10	19	=	7
20	=	12	20	=	14	20	=	11	20	=	9	20	=	19	20	=	18	20	=	9
21	=	5	21	=	15	21	=	5	21	=	9	21	=	10	21	=	20	21	=	13
22	=	16	22	=	13	22	=	6	22	=	9	22	=	12	22	=	11	22	=	6
23	=	20	23	=	10	23	=	10	23	=	8	23	=	9	23	=	9	23	=	9
24	=	16	24	=	4	24	=	1	24	=	6	24	=	10	24	=	7	24	=	10
25	=	16	25	=	10	25	=	19	25	=	10	25	=	2	25	=	16	25	=	2
26	=	7	26	=	5	26	=	10	26	=	14	26	=	6	26	=	17	26	=	7
27	=	5	27	=	1	27	=	13	27	=	6	27	=	14	27	=	9	27	=	12
28	=	3	28	=	7	28	=	2	28	=	13	28	=	9	28	=	10	28	=	6
29	=	15	29	=	14	29	=	18	29	=	10	29	=	12	29	=	12	29	=	11
30	=	12	30	=	6	30	=	14	30	=	8	30	=	16	30	=	13	30	=	7
31	=	3	31	=	6	31	=	12	31	=	8	31	=	12	31	=	4	31	=	14
32	=	13	32	=	16	32	=	10	32	=	7	32	=	15	32	=	11	32	=	11
33	=	11	33	=	16	33	=	6	33	=	8	33	=	11	33	=	18	33	=	5
34	=	11	34	=	2	34	=	15	34	=	11	34	=	8	34	=	4	34	=	11
35	=	13	35	=	2	35	=	3	35	=	5	35	=	8	35	=	12	35	=	13
36	=	15	36	=	13	36	=	6	36	=	12	36	=	13	36	=	8	36	=	9

Day 36	Day 37	Day 38	Day 39	Day 40	Day 41	Day 42
1 = 5	1 = 0	1 = 9	1 = 2	1 = -3	1 = -3	1 = -7
2 = 1	2 = 1	2 = 2	2 = 1	2 = 4	2 = -1	2 = -1
3 = -2	3 = -2	3 = 0	3 = 2	3 = -4	3 = 5	3 = -3
4 = -5	4 = 1	4 = 10	4 = 0	4 = 0	4 = 5	4 = 1
5 = 0	5 = -4	5 = 1	5 = -2	5 = 5	5 = -2	5 = 0
6 = -5	6 = 1	6 = -4	6 = 1	6 = -1	6 = -1	6 = -8
7 = 8	7 = -3	7 = 2	7 = -3	7 = 0	7 = 3	7 = -1
8 = 1	8 = -2	8 = -1	8 = -4	8 = -3	8 = 7	8 = 6
9 = 6	9 = -2	9 = 2	9 = 10	9 = 3	9 = -8	9 = -4
10 = 3	10 = 2	10 = -8	10 = -2	10 = -7	10 = -2	10 = 6
11 = 6	11 = 3	11 = -3	11 = 2	11 = -3	11 = -3	11 = -4
12 = -2	12 = 1	12 = 8	12 = -8	12 = -4	12 = 10	12 = -8
13 = -2	13 = 1	13 = 7	13 = -3	13 = 0	13 = 4	13 = -8
14 = 0	14 = 4	14 = -3	14 = -2	14 = 4	14 = 2	14 = -1
15 = 3	15 = 0	15 = 0	15 = 3	15 = 2	15 = 2	15 = -2
16 = 3	16 = -3	16 = -10	16 = -5	16 = 2	16 = -5	16 = 0
17 = 4	17 = 9	17 = -6	17 = -2	17 = 8	17 = 2	17 = 6
18 = -3	18 = 4	18 = 4	18 = 0	18 = 1	18 = 1	18 = -5
19 = 1	19 = 8	19 = -5	19 = 4	19 = -5	19 = 1	19 = 1
20 = 1	20 = 1	20 = -6	20 = 1	20 = -5	20 = 0	20 = 2
21 = -3	21 = -1	21 = 9	21 = 7	21 = -2	21 = -1	21 = 9
22 = 7	22 = 3	22 = 7	22 = -3	22 = -2	22 = -6	22 = 9
23 = 1	23 = 4	23 = -2	23 = -2	23 = -1	23 = -7	23 = -3
24 = 4	24 = -9	24 = 6	24 = -7	24 = -6	24 = 8	24 = -2
25 = 1	25 = 0	25 = -2	25 = 1	25 = 4	25 = 4	25 = 10
26 = 4	26 = 2	26 = -4	26 = 10	26 = -3	26 = -1	26 = 2
27 = 0	27 = -1	27 = 1	27 = -1	27 = -3	27 = 2	27 = -5
28 = -9	28 = 3	28 = -4	28 = 4	28 = 0	28 = -1	28 = 0
29 = 9	29 = -1	29 = 1	29 = -5	29 = 4	29 = -2	29 = 1
30 = 10	30 = -2	30 = 7	30 = 3	30 = 8	30 = 6	30 = 3
31 = -8	31 = -3	31 = 3	31 = -2	31 = 0	31 = -1	31 = 3
32 = -2	32 = -7	32 = 0	32 = 3	32 = 3	32 = -4	32 = -4
33 = -2	33 = -10	33 = -8	33 = 7	33 = 6	33 = -2	33 = 9
34 = 5	34 = 3	34 = 2	34 = -2	34 = -3	34 = -10	34 = 0
35 = -5	35 = 0	35 = -2	35 = 3	35 = -1	35 = 4	35 = 7
36 = 3	36 = -8	36 = -3	36 = -3	36 = 2	36 = -2	36 = -2

Day 43	Day 44	Day 45	Day 46	Day 47	Day 48	Day 49
1 = -5	1 = -4	1 = -6	1 = 5	1 = -4	1 = -5	1 = -6
2 = -2	2 = -5	2 = -6	2 = 3	2 = -1	2 = 1	2 = 0
3 = 2	3 = -2	3 = 0	3 = 0	3 = -9	3 = 2	3 = 9
4 = 4	4 = 2	4 = 5	4 = -4	4 = -3	4 = 0	4 = 2
5 = 4	5 = -5	5 = -5	5 = -7	5 = 0	5 = 0	5 = 6
6 = 10	6 = 2	6 = -3	6 = -3	6 = 0	6 = 2	6 = 0
7 = -1	7 = 6	7 = -6	7 = -2	7 = 8	7 = 5	7 = 10
8 = -2	8 = -7	8 = -8	8 = -5	8 = 7	8 = 0	8 = -2
9 = 1	9 = 6	9 = 1	9 = 7	9 = 0	9 = 4	9 = -6
10 = -4	10 = 1	10 = -5	10 = 0	10 = 6	10 = 8	10 = 2
11 = -4	11 = 4	11 = -3	11 = -6	11 = 4	11 = 1	11 = -4
12 = 4	12 = -5	12 = 8	12 = -1	12 = 1	12 = -2	12 = 9
13 = 0	13 = 3	13 = -3	13 = -5	13 = 0	13 = -4	13 = -2
14 = 2	14 = 3	14 = 8	14 = 1	14 = -1	14 = -4	14 = -1
15 = 5	15 = -2	15 = -5	15 = 4	15 = 6	15 = 8	15 = 2
16 = 6	16 = 1	16 = -4	16 = -3	16 = -9	16 = 0	16 = -2
17 = -8	17 = -1	17 = -2	17 = -2	17 = -8	17 = -9	17 = 3
18 = 6	18 = 1	18 = 1	18 = 6	18 = -3	18 = 2	18 = -3
19 = -2	19 = -5	19 = 3	19 = -1	19 = -9	19 = 6	19 = 0
20 = -4	20 = -7	20 = 2	20 = 5	20 = -6	20 = 4	20 = 7
21 = -5	21 = -2	21 = -7	21 = 1	21 = 3	21 = 3	21 = 1
22 = 0	22 = 2	22 = 5	22 = -1	22 = -5	22 = 1	22 = 5
23 = -4	23 = 7	23 = 4	23 = -2	23 = -1	23 = -6	23 = 1
24 = 7	24 = 7	24 = -2	24 = 7	24 = 4	24 = -1	24 = 0
25 = -1	25 = 0	25 = 0	25 = 0	25 = -2	25 = -6	25 = -1
26 = -4	26 = 1	26 = 1	26 = 0	26 = 10	26 = 1	26 = 7
27 = 7	27 = 2	27 = 1	27 = -3	27 = -4	27 = 6	27 = -3
28 = 3	28 = 2	28 = 1	28 = 6	28 = 2	28 = -4	28 = 0
29 = 8	29 = 3	29 = -7	29 = -1	29 = -6	29 = -5	29 = 8
30 = -3	30 = 10	30 = -3	30 = 7	30 = -3	30 = 3	30 = -8
31 = 1	31 = 8	31 = 1	31 = 0	31 = -2	31 = -2	31 = 1
32 = 4	32 = 1	32 = 2	32 = 9	32 = -1	32 = -6	32 = -5
33 = 0	33 = -5	33 = 5	33 = 8	33 = 4	33 = -4	33 = 1
34 = 10	34 = -1	34 = -2	34 = -2	34 = 2	34 = 6	34 = 3
35 = 4	35 = -6	35 = -1	35 = 6	35 = 6	35 = 0	35 = -2
36 = 2	36 = 6	36 = -1	36 = 0	36 = -2	36 = -4	36 = 2

Day 50	Day 51	Day 52	Day 53	Day 54	Day 55	Day 56
1 = -7	1 = 6	1 = 0	1 = -2	1 = 0	1 = 2	1 = -6
2 = -8	2 = -6	2 = -10	2 = 3	2 = 3	2 = -2	2 = 3
3 = 4	3 = -3	3 = -4	3 = -6	3 = -5	3 = 1	3 = 0
4 = -2	4 = 1	4 = -4	4 = -3	4 = -3	4 = -2	4 = 7
5 = -4	5 = 0	5 = 7	5 = -6	5 = -7	5 = -5	5 = -5
6 = 2	6 = -8	6 = 2	6 = -5	6 = -2	6 = 0	6 = 4
7 = -8	7 = 0	7 = 6	7 = 1	7 = -2	7 = -5	7 = -7
8 = -9	8 = -5	8 = 2	8 = -5	8 = -2	8 = -3	8 = -1
9 = 7	9 = 1	9 = 2	9 = -5	9 = 0	9 = -1	9 = 4
10 = 4	10 = 1	10 = -4	10 = -3	10 = -1	10 = 9	10 = -5
11 = -2	11 = -1	11 = 2	11 = 3	11 = 1	11 = 9	11 = -1
12 = -3	12 = 0	12 = 9	12 = 10	12 = 7	12 = 3	12 = 6
13 = 3	13 = 3	13 = 3	13 = 1	13 = -1	13 = 5	13 = 2
14 = 2	14 = -5	14 = -2	14 = 7	14 = -4	14 = 0	14 = 3
15 = -6	15 = 5	15 = -5	15 = 4	15 = -3	15 = 2	15 = -4
16 = 1	16 = 2	16 = 1	16 = -4	16 = 2	16 = -4	16 = 8
17 = 6	17 = 5	17 = 0	17 = -8	17 = -7	17 = 0	17 = -1
18 = 5	18 = 1	18 = -1	18 = -3	18 = 7	18 = -1	18 = -5
19 = 3	19 = 6	19 = -4	19 = -1	19 = 2	19 = -4	19 = 0
20 = 0	20 = 7	20 = -6	20 = 4	20 = -6	20 = 2	20 = 0
21 = 3	21 = 0	21 = -1	21 = 0	21 = 0	21 = -4	21 = -3
22 = 10	22 = 2	22 = -2	22 = 6	22 = -7	22 = -2	22 = -2
23 = -6	23 = -4	23 = 0	23 = 6	23 = -2	23 = -5	23 = 0
24 = -2	24 = 4	24 = -3	24 = -2	24 = 5	24 = -3	24 = 1
25 = 0	25 = 1	25 = -2	25 = -2	25 = 0	25 = 4	25 = 9
26 = -2	26 = -6	26 = 1	26 = 0	26 = 0	26 = -4	26 = 2
27 = -4	27 = -6	27 = -2	27 = -4	27 = 2	27 = 6	27 = 4
28 = 5	28 = 0	28 = -5	28 = 8	28 = -8	28 = -6	28 = -7
29 = 2	29 = 1	29 = -5	29 = 0	29 = -6	29 = -1	29 = -5
30 = 0	30 = -4	30 = 2	30 = -4	30 = -2	30 = 1	30 = -1
31 = -7	31 = -3	31 = 0	31 = -1	31 = -3	31 = 3	31 = -4
32 = 4	32 = -5	32 = -1	32 = 1	32 = 5	32 = 2	32 = -5
33 = 0	33 = 7	33 = 3	33 = -7	33 = -2	33 = -1	33 = -9
34 = 1	34 = 3	34 = -1	34 = -2	34 = -3	34 = 2	34 = -1
35 = 0	35 = -1	35 = 7	35 = -5	35 = -2	35 = -3	35 = 0
36 = -5	36 = 4	36 = 2	36 = -1	36 = 2	36 = 5	36 = -2

Day 57	Day 58	Day 59	Day 60	Day 61	Day 62	Day 63
1 = 4	1 = -5	1 = -1	1 = -3	1 = 19	1 = -5	1 = 8
2 = -15	2 = -1	2 = 1	2 = -12	2 = 5	2 = -8	2 = -18
3 = 6	3 = 1	3 = 13	3 = -4	3 = -7	3 = 8	3 = 3
4 = -8	4 = -13	4 = 16	4 = -7	4 = -4	4 = 13	4 = -2
5 = 3	5 = 0	5 = 1	5 = -7	5 = 1	5 = 6	5 = 1
6 = -1	6 = 2	6 = -10	6 = -5	6 = -7	6 = 4	6 = -16
7 = 15	7 = 6	7 = 8	7 = 0	7 = 3	7 = -12	7 = -2
8 = 2	8 = -16	8 = 6	8 = -4	8 = -8	8 = -3	8 = 10
9 = -6	9 = -13	9 = -8	9 = -1	9 = 3	9 = -5	9 = 1
10 = 2	10 = 0	10 = -1	10 = -8	10 = -5	10 = -1	10 = 5
11 = 9	11 = 1	11 = -5	11 = 8	11 = 2	11 = 6	11 = 6
12 = -10	12 = 0	12 = -4	12 = -5	12 = -9	12 = -8	12 = -4
13 = -18	13 = -7	13 = 6	13 = 7	13 = -8	13 = 5	13 = 4
14 = -7	14 = -4	14 = 14	14 = -9	14 = 11	14 = 9	14 = 1
15 = 0	15 = 2	15 = -14	15 = 4	15 = 19	15 = 1	15 = 9
16 = -2	16 = -1	16 = -9	16 = 6	16 = -11	16 = 1	16 = 17
17 = -12	17 = -4	17 = -8	17 = 8	17 = -11	17 = 11	17 = 3
18 = 0	18 = 3	18 = 4	18 = -6	18 = -18	18 = -7	18 = -8
19 = -8	19 = 9	19 = -12	19 = -15	19 = -16	19 = -6	19 = 2
20 = 2	20 = 7	20 = -6	20 = 15	20 = 1	20 = 0	20 = -2
21 = -1	21 = 1	21 = -12	21 = -3	21 = -7	21 = 11	21 = 10
22 = 1	22 = 0	22 = -5	22 = 10	22 = -20	22 = 18	22 = -1
23 = 1	23 = -5	23 = -9	23 = -3	23 = 6	23 = 10	23 = 16
24 = 1	24 = 12	24 = -7	24 = 6	24 = 0	24 = 12	24 = 13
25 = -18	25 = 19	25 = -1	25 = -15	25 = 0	25 = 0	25 = -7
26 = -13	26 = 6	26 = 4	26 = 5	26 = -13	26 = -6	26 = -4
27 = 4	27 = -9	27 = 14	27 = 3	27 = 1	27 = 12	27 = 5
28 = 13	28 = -9	28 = 5	28 = 14	28 = 4	28 = 17	28 = 16
29 = -3	29 = -13	29 = 5	29 = -15	29 = -15	29 = 3	29 = 1
30 = 12	30 = -6	30 = 4	30 = 5	30 = 12	30 = 6	30 = -2
31 = 0	31 = 2	31 = -7	31 = -1	31 = 2	31 = 14	31 = 5
32 = -5	32 = 1	32 = 14	32 = -2	32 = -2	32 = 0	32 = -6
33 = 6	33 = -5	33 = -2	33 = 8	33 = 5	33 = -8	33 = -13
34 = -1	34 = -1	34 = -11	34 = 5	34 = -8	34 = -14	34 = 3
35 = 6	35 = -4	35 = 4	35 = 3	35 = -10	35 = 3	35 = -9
36 = 5	36 = 12	36 = 15	36 = -17	36 = 7	36 = 14	36 = -17

Day 64	Day 65	Day 66	Day 67	Day 68	Day 69	Day 70
1 = 11	1 = -1	1 = 16	1 = -6	1 = 1	1 = 6	1 = -2
2 = -3	2 = -5	2 = -2	2 = 16	2 = 7	2 = 7	2 = 16
3 = 9	3 = -16	3 = -8	3 = 8	3 = 2	3 = 6	3 = 16
4 = 7	4 = 15	4 = -9	4 = 0	4 = -9	4 = -7	4 = -7
5 = -12	5 = 9	5 = -4	5 = 14	5 = -3	5 = -13	5 = -4
6 = -13	6 = -8	6 = 12	6 = 8	6 = 6	6 = 10	6 = -5
7 = 8	7 = -14	7 = 14	7 = 10	7 = 15	7 = -6	7 = -2
8 = -9	8 = -2	8 = 16	8 = 0	8 = 0	8 = -8	8 = 7
9 = -2	9 = 6	9 = -1	9 = -10	9 = -5	9 = 3	9 = -5
10 = 11	10 = -4	10 = 8	10 = 3	10 = -8	10 = 13	10 = 2
11 = 8	11 = -8	11 = 5	11 = 6	11 = -7	11 = 1	11 = 6
12 = -11	12 = -13	12 = 1	12 = 6	12 = -3	12 = -3	12 = 0
13 = 4	13 = 9	13 = -1	13 = -13	13 = 11	13 = 11	13 = -19
14 = -4	14 = 9	14 = 9	14 = -4	14 = -14	14 = 8	14 = -7
15 = -6	15 = 3	15 = -6	15 = 18	15 = -4	15 = 10	15 = 6
16 = -2	16 = -2	16 = 5	16 = 3	16 = 1	16 = 4	16 = 2
17 = -2	17 = -16	17 = 8	17 = 0	17 = -8	17 = -4	17 = -12
18 = 5	18 = 3	18 = -10	18 = 18	18 = 3	18 = -7	18 = 6
19 = 12	19 = 15	19 = 12	19 = 4	19 = 10	19 = 7	19 = 7
20 = -4	20 = -13	20 = -1	20 = 13	20 = 4	20 = 7	20 = 15
21 = 5	21 = -9	21 = -4	21 = 8	21 = 10	21 = -10	21 = -3
22 = -6	22 = 5	22 = 6	22 = 9	22 = -7	22 = 16	22 = -8
23 = 9	23 = -16	23 = 8	23 = -7	23 = -1	23 = 11	23 = 3
24 = -15	24 = -3	24 = -1	24 = -1	24 = 8	24 = 3	24 = -7
25 = 10	25 = 11	25 = -6	25 = 12	25 = 11	25 = 16	25 = 13
26 = 15	26 = -9	26 = -7	26 = -16	26 = -7	26 = 3	26 = -18
27 = 19	27 = -12	27 = 17	27 = 12	27 = 10	27 = -5	27 = 12
28 = 4	28 = -1	28 = -17	28 = -1	28 = -1	28 = -16	28 = 13
29 = 14	29 = -10	29 = 13	29 = 7	29 = -8	29 = -2	29 = -3
30 = -3	30 = 1	30 = 13	30 = 6	30 = -15	30 = 10	30 = 3
31 = -5	31 = 0	31 = -20	31 = 6	31 = 5	31 = 4	31 = 7
32 = -6	32 = -1	32 = 1	32 = 9	32 = -4	32 = -5	32 = 12
33 = -7	33 = -9	33 = -10	33 = -4	33 = -7	33 = 0	33 = 3
34 = 6	34 = 7	34 = -10	34 = 3	34 = -5	34 = -3	34 = 3
35 = 6	35 = 2	35 = 5	35 = -2	35 = -8	35 = 12	35 = 0
36 = 6	36 = -9	36 = 7	36 = 0	36 = -11	36 = 10	36 = 8

www.ingramcontent.com/pod-product-compliance
Lightning Source LLC
Chambersburg PA
CBHW081057240526
45465CB00025B/2503